edZOOcation
SLOTHS

ZOOLOGIST AGES 6-8

Wildlife Tree
edZOOcation

Dedication:

Melissa, a motivation in my design journey.

–A.R.

Curtis, Jenny. (2024). Assisted by OpenAI's ChatGPT

Copyright © 2023 Wildlife Tree, LLC. All rights reserved.

Designer: Allyson Randa

Photo Credits:

AdobeStock.com

Pixabay.com

Pexels.com

ISBN: 979-8-9898789-3-2

This book meets **Common Core** and **Next Generation Science** Standards.

Table of Contents

4	Meet the Sloth
6	A Sloth's Body
8	Special Skills
10	Poop
11	Species
14	Hidden Sloths
15	Sloth Babies
17	High Rise Homes
18	Where Are They?
20	Where Is Home?
21	Sleepy Sloths
22	A Light Lunch
24	Food Web
25	Dangers
26	Glossary

Meet the Sloth

Have you ever seen a sloth? They may seem slow and sleepy, but there's so much more to learn about these amazing tree-huggers.

Did you know that sloths can't see well in bright light, but have a super sense of smell? And did you know that baby sloths change colors as they grow up?

Let's dive into their secret world together!

A Sloth's Body

Small eyes

Mouths that always look like they're smiling

Algae: *Simple, plant-like life-forms, usually found in water.*

Long claws

Long arms for hanging

Fur that grows *algae*

Special Skills

Sloths can hang upside down all day, thanks to their special hands and feet. Their muscles lock into place so that they can even sleep upside down. Their grip is so powerful, even a jaguar may not be able to rip them from a tree!

Sloths are great swimmers too. Even though they're slow on land, in the water, they're three times as fast.

Poop

Sloths only poop on the ground. Going down to the ground is dangerous since they move so slowly. So, they only go down to poop once a week. That's a long time to hold it!

Species

Sloths are amazing creatures that you can tell apart by looking at their claws! There are two main types: The first are two-toed sloths, who have two claws on their front feet. The second are three-toed sloths, who have three claws on all their feet.

Species

Two-toed sloths are a bit bigger and like to hang upside down more. Three-toed sloths often sit up in tree branches.

There are only two different species of two-toed sloths. There are four different species of three-toed sloths.

Hidden Sloths

Sloths are great at hiding. Can you spot the 5 sloths?

Sloth Babies

Baby sloths are born ready to cling to their mom's fur for safety and milk. They start eating leaves like their mom when they're just one week old. As they grow, they learn to grab branches and explore more, but always stay close to mom.

After about six months, they start to be more **independent**, but still hang out near her. Sloth moms are great teachers, helping their babies learn what to eat and how to live in the forest.

Independent: *Not needing help from others.*

High-Rise Home

Sloths live high up in the trees of the rainforest. You might think that living up so high would be dangerous. But sloths have a special superpower – they can fall 100 feet without getting hurt. That's as long as an average passenger airplane!

Where Are They?

Sloths live in the tropical rainforests of South and Central America.

Where is Home?

In which of these places would you find a sloth?

Rainforest　　Desert　　Ocean

Central America　　Prairie　　South America

Answer key: Rainforest, Central America, South America

Sleepy Sloths

Three-fingered sloths might be awake both day and night, but two-fingered sloths usually wake up at night. Studying their sleep is hard, but researchers found that sloths sleep or rest a lot. Each sloth sleeps differently, which might help them avoid **predators**. Even though sloths are known to be sleepy, they usually only sleep for 15 hours a day or less. They also spend several hours a day resting.

Predator: *An animal that hunts other animals.*

A Light Lunch

Sloths love to munch on leaves, and fresh new leaves are their favorites. A leaf will last a long time, though. It takes them 30 days to digest a leaf!

Sometimes, two-fingered sloths will eat dirt.

Digest: *Break down food that's been eaten.*

Food Web

Sloths only eat leaves – they don't eat other animals. But they are hunted by ocelots, eagles, and jaguars.

Dangers

Some sloths are in danger because their homes in the forest are disappearing. But many people are working hard to protect them and their homes. We can help sloths by learning more about them and sharing their story!

Glossary

Algae: Simple, plant-like life-forms, usually found in water.

Independent: Not needing help from others.

Digest: Break down food that's been eaten.

Predator: An animal that hunts other animals.

Silly Sloths

In treetops high, where dreams softly float,
Dwells the gentle sloth, in a leafy green coat.
With slow, graceful moves, through branches they roam.
In the lush rainforest, they make their home.

Q: What did the sloth say to its friend?

A: "Hang in there, buddy!"

Q: How do sloths communicate online?

A: They use slow-cial media.

Q: When does a sloth go "moo?"

A: When it's learning a new language!

Q: Ever heard the one about the sloth crossing the road?

A: Never mind, it'd take too long.

Q: What's a sloth's favorite song?

A: "Don't Hurry, Be Happy."

Q: Why did the sloth cross the road?

A: Nobody knows, he's still trying.